TREASURE ISLANDS

'And the spirit of all that I beheld put me in thoughts of far voyages and foreign places.'

TREASURE ISLANDS

A Robert Louis Stevenson Centenary Anthology
compiled and edited by Jenni Calder

N·M·S

NATIONAL MUSEUMS OF SCOTLAND
Sponsored by Martin Currie Investment Management Ltd

Front cover: Pirates attacking a galleon. *Illustration by Howard Pyle.*
Back cover: Boys Sailing Boats *by William Marshall Brown.*
Frontispiece: At Queen's Ferry. *Illustration for* Kidnapped *by N C Wyeth.*

The centenary of the death of Robert Louis Stevenson on 3 December 1894 was marked by the exhibition 'Treasure Islands', at the Royal Museum of Scotland, August 1994 to January 1995.

Acknowledgements
Grateful thanks to NMS staff and the following for help with pictures and picture research: Dorota Starzecka and Harry Persaud of the Museum of Mankind; Dr Jo Ann Van Tilburg, University of California; Elaine Finnie, The Writers' Museum, Edinburgh; Dr Merrily Stover, Washington DC; Embassy of Western Samoa, Brussels; Baxter Nisbet, Lochgilphead; Arthur Blue, Ardrishaig.

Thanks for help and advice to Dale Idiens and Maureen Barrie, Department of History and Applied Art, NMS

Maps by Steve Gibson

Illustration acknowledgements for:
Front cover, 11, 46, 71: National Maritime Museum London. 2: Reprinted with the permission of Charles Scribner's Sons, an imprint of Macmillan Publishing Company, from *Kidnapped* by Robert Louis Stevenson, illustrated by N C Wyeth. Copyright 1913 Charles Scribner's Sons; copyright renewed 1941 N C Wyeth. 3, 9, 23, 37, 41, 43, 45: Trustees of the National Library of Scotland. 5, 15, 16/17, 22, 33, 36, 54, 55(bottom), 60, 62, 73, 80, 81, 85: Trustees of the National Museums of Scotland. 7: © Photo R.M.N. 10, 64: by courtesy of Edinburgh City Libraries. 12, 53, 61, 65: Edinburgh City Museums, The Writers' Museum. 18: Glasgow Museums: Art Gallery & Museum, Kelvingrove. 20, 27, 30, 77, 86: National Gallery of Scotland. 28/29 Edinburgh University Library. 32: SNPG, Department of Photography. 39, 67: The Bridgeman Art Library, London. 47: illustration by Gustav Doré from the *Rime of the Ancient Mariner*, courtesy Dover Publications, Inc. 51: photograph courtesy of the Lefevre Gallery, London. 75: *Te raau rahi (The Big Tree).* Oil on canvas, 1891, 74 x 92.8 cm. Paul Gauguin, France, 1848-1903. © The Cleveland Museum of Art, gift of Barbara Ginn Griesïnger, 75.263. 82: Picturepoint-London. Back cover: Bourne Fine Art, Edinburgh.

Published by the National Museums of Scotland,
Chambers Street,
Edinburgh EH1 1JF

ISBN 0 948636 59 9

Designed and produced by the
Publications Office of the National Museums of Scotland
Printed by Craftprint (Singapore) Ltd

© Trustees of the National Museums of Scotland 1994

British Library Cataloguing in Publication Data
A catalogue record for this book is available from the British Library

To *The Hesitating Purchaser*

If sailor tales to sailor tunes,
    Storm and adventure, heat and cold,
If schooners, islands, and maroons
    And Buccaneers and buried Gold,
And all the old romance, retold
    Exactly in the ancient way,
Can please, as me they pleased of old,
    The wiser youngsters of to-day:

– So be it, and fall on! If not,
    If studious youth no longer crave,
His ancient appetites forgot,
    Kingston, or Ballantyne the brave,
Or Cooper of the wood and wave:
    So be it, also! And may I
And all my pirates share the grave
    Where these and their creations lie!

Treasure Island

ISLANDS OF ADVENTURE

This book is a collection of islands. Big islands and small islands, dangerous islands and magic islands. Islands with rocks and ruins, islands with beaches and palm trees. Cold Scottish islands and sunny islands in the Pacific Ocean. Real islands and islands of the imagination. They come from the books of Robert Louis Stevenson, who from his childhood explored islands and as an adult wrote about them.

As a child in Victorian Edinburgh, RLS, as he is often called, heard many stories of islands and rocky shores, rough seas and dangerous voyages. He came from a family of engineers who built most of Scotland's best-known lighthouses. He was fascinated by the sea and ships, and everything connected with them.

Naufrage *from Gauguin's album* Noa Noa. *The artist Gauguin was in Tahiti at the same time as RLS was in Samoa.* Louvre

Born in 1850, from the age of seven he lived in Heriot Row, opposite Queen Street Gardens where there was a small pond with an island. The pond in nearby Inverleith Park was even better for sailing model boats. Sometimes he was taken for walks down to Leith, where he watched the ships in the harbour and wondered where they had all come from.

Opposite: *Treshnish Isles, West Mull.* Charles Tait

His grandfather was minister at Colinton, a village outside Edinburgh. When the young RLS went to stay he played by the Water of Leith and sent paper boats downstream. He was an only child, but at Colinton he spent time with his cousins. It was an ideal place to play at explorers and soldiers, to be a ship's captain or a pirate, a hunter or an Indian scout.

He spent several summer holidays in North Berwick, with the Bass Rock a short boat trip away. He explored the beach, the cliffs and rock pools, and Tantallon Castle nearby.

But often he was ill, and had to be content with making up stories of travel and adventure. He invented his own oceans and islands when he wasn't able to get out to the real thing. His illnesses continued when he was grown up, and eventually he had to leave Scotland and live in a warmer climate.

He knew the islands in the Firth of Forth - Cramond, Inchkeith, Inchcolm and others. When he was older, he went with his father on trips to inspect lighthouses round Scotland's coasts, in Shetland, Orkney and the Hebridean islands.

His love of islands and the sea became a part of many of the stories and poems RLS wrote. He began writing when he was very young, but his father wanted him to follow in his footsteps and become a lighthouse engineer. He studied engineering at Edinburgh University, until he finally told his parents he wanted to write for a living.

His most famous island story is *Treasure Island*, about young Jim Hawkins of the Admiral Benbow inn, who joins a group of

Jim Hawkins with Long John Silver, on Bristol quay. An illustration by Wal Paget from Treasure Island, *1899.*

RLS and his wife Fanny in Butaritari, Kiribati (the Gilbert Islands) with Nan Tok' and Nei Takauti.

adventurers on the schooner *Hispaniola* to search for hidden treasure. They encounter pirates, including Long John Silver, one of RLS's most famous characters, and have many mishaps before eventually succeeding in their quest.

In search of better health, RLS tried living in Switzerland, the south of France, the south of England and America. He could never stay in Scotland for long. He had always wanted to sail the Pacific and visit the islands of the South Seas. Finally, in 1888, he was able to charter a schooner yacht, the *Casco*, and fulfil his dream. His mother, his American wife Fanny, and his step-son Lloyd sailed with him.

For two years he voyaged from island to island, in the *Casco* and two other ships, the *Equator*, a schooner, and the *Janet Nichol*, a trading steamer. He made many friends among the island peoples and was

Opposite: RLS loved pirate stories. Buried Treasure. *Illustration by Howard Pyle.*

RLS and his family at Vailima, his Samoan home. RLS and Fanny sit side by side with his mother standing on his left. On the steps sit Fanny's children Lloyd and Belle, with Belle's son Austin and her husband Joe Strong standing, a cockatoo on his shoulder.

Right: *Vailima. Watercolour by Belle, 1891.*

deeply interested in their way of life, their songs and stories, and the objects they made. He protested at the way their life had been affected, in his view for the worse, by European traders and missionaries who came to the islands.

The South Seas suited his health, and he decided to stay there. He built a house, Vailima, on the island of Upolu in Samoa. He and Fanny worked hard to clear the ground and grow fruit and vegetables to feed the household. But the most important thing was to carry on with his writing. He produced many more books, about both the South Seas and Scotland. He was also a great letter-writer. The Samoans had great respect for him and called him 'Tusitala' - teller of tales. It was at Vailima that he died on 3 December 1894.

By the time of his death, his books were being enjoyed all over the world, by readers of all ages. They have been translated into many languages. He is now recognized everywhere as one of Scotland's best and most important writers.

The islands in this collection are found in Stevenson's poems, stories, novels, essays and letters. They offer a taste of RLS. They show him as a writer of travel and adventure, and also as an explorer of the realms of the imagination. They show children at play and heroes and villains in action. They bring vividly to life the world of ships and oceans. Most of all, we hope they will tempt the reader to ask for more, and go to the books from which the extracts come.

RLS was an only child, and often ill. To amuse himself, he invented many games - games of adventure, travel, ships and foreign lands.

from Travel

A market in Apia, Western Samoa, today. The Stevenson family brought their supplies on horseback from Apia. Merrily Stover

I should like to rise and go
Where the golden apples grow; -
Where below another sky
Parrot islands anchored lie,
And, watched by cockatoos and goats,
Lonely Crusoes building boats;-
Where in sunshine reaching out
Eastern cities, miles about,
Are with mosque and minaret
Among sandy gardens set,
And the rich goods from near and far
Hang for sale in the bazaar...

Pirate Story

Three of us afloat in the meadow by the swing,
    Three of us aboard in the basket on the lea.
Winds are in the air, they are blowing in the spring,
    And waves are on the meadow like the waves there
           are at sea.

Where shall we adventure, to-day that we're afloat,
    Wary of the weather and steering by a star?
Shall it be to Africa, a-steering of the boat,
    To Providence, or Babylon, or off to Malabar?

Hi! but here's a squadron a-rowing on the sea –
    Cattle on the meadow a-charging with a roar!
Quick, and we'll escape them, they're as mad as they can be,
    The wicket is the harbour and the garden is
           the shore.

Treasure chest, from the collections of the
National Museums of Scotland

Block City

What are you able to build with your blocks?
Castles and palaces, temples and docks.
Rain may keep raining, and others go roam,
But I can be happy and building at home.

Let the sofa be mountains, the carpet be sea,
There I'll establish a city for me:
A kirk and a mill and a palace beside,
And a harbour as well where my vessels may ride.

Great is the palace with pillar and wall,
A sort of a tower on the top of it all,
And steps coming down in an orderly way
To where my toy vessels lie safe in the bay.

Panorama of the city of Edinburgh looking to the west from Calton Hill, from Daniel Wilson's scrapbook.

This one is sailing and that one is moored:
Hark to the song of the sailors on board!
And see on the steps of my palace, the kings
Coming and going with presents and things!

Now I have done with with it, down let it go!
All in a moment the town is laid low.
Block upon block lying scattered and free,
What is there left of my town by the sea?

Yet as I saw it, I'll see it again,
The kirk and the palace, the ship and the men,
And as long as I live and where'er I may be,
I'll always remember my town by the sea.

On holiday at North Berwick, the young Louis, as he was called, played on the shore opposite the Bass Rock, then, as now, the home of thousands of gannets (solan geese).

from The Lantern-Bearers

A street or two of houses, mostly red and many of them tiled; a number of fine trees clustered about the manse and the kirk-yard, and turning the chief street into a shady alley; many little gardens more than usually bright with flowers; nets a-drying, and fisher-wives scolding in the backward parts; a smell of fish, a genial smell of seaweed; whiffs of blowing sand at the street-corners; shops with gold-balls and bottled lollipops...a haven in the rocks in

Evening, North Berwick, by S J Peploe, about 1905.

*Islands of RLS's childhood. 'A file of gray islets' and the
Bass Rock.* Don McKinnell

front: in front of that a file of gray islets: to the left, endless links and
sand wreaths, a wilderness of hiding-holes, alive with popping
rabbits and soaring gulls: to the right, a range of seaward crags, one
rugged brow beyond another; the ruins of a mighty and ancient
fortress on the brink of one; coves between – now charmed into
sunshine quiet, now whistling with wind and clamorous with
bursting surges; the dens and sheltered hollows redolent of thyme
and southernwood, the air at the cliff's edge brisk and clean and
pungent of the sea – in front of all the Bass Rock, tilted seaward like
a doubtful bather, the surf ringing it with white, the solan-geese
hanging round its summit like a great and glittering smoke...

...you might climb the Law, where the whale's jawbone stood landmark in the buzzing wind, and behold the face of many counties, and the smoke and spires of many towns, and the sails of distant ships. You might bathe, now in the flaws of fine weather, that we pathetically call our summer, now in a gale of wind, with the sand scouring your bare hide, your clothes thrashing abroad from underneath their guardian stone, the froth of the great breakers casting you headlong ere it had drowned your knees. Or you might explore the tidal rocks, above all in the ebb of springs, when the very roots of the hills were for the nonce discovered; following my leader from one group to another, groping in slippery tangle for the wreck of ships, wading in pools after the abominable creatures of the sea, and ever with an eye cast backward on the march of the tide and the menaced line of your retreat.

'Just to the Bass, mannie.' Boats off the Bass Rock,
by Alexander Nasmyth, 1820.

When RLS wrote *Catriona*, he set part of the story on the Bass Rock. David Balfour, the hero, is captured and held prisoner on the Rock, where in the seventeenth century Covenanters - 'the auld sants' - had been imprisoned.

from Catriona

There began to fall a greyness on the face of the sea; little dabs of pink and red, like coals of slow fire, came in the east; and at the same time the geese awakened, and began crying about the top of the Bass. It is just the one crag of rock, as everybody knows, but great enough to carve a city from. The sea was extremely little, but there went a hollow plowter round the base of it. With the growing of the dawn I could see it clearer and clearer; the straight crags painted with sea-birds' droppings like a morning frost, the sloping top of it green with grass, the clan of white geese that cried about the sides, and the black, broken buildings of the prison sitting close on the sea's edge.

At the sight the truth came in upon me in a clap.

'It's there you're taking me!' I cried.

'Just to the Bass, mannie,' said he: 'Whaur the auld sants went afore ye, and I misdoubt if ye had come so fairly by your preeson.'

'But none dwells there now,' I cried; 'the place is long a ruin.'

'It'll be the mair pleisand a change for the solan geese, then,' quoth Andie dryly.

Gannets (solan geese), on display at the Royal Museum of Scotland, 'their droppings like a morning frost'.

While a prisoner on the Rock, David's guard Andie tells him 'The Tale of Tod Lapraik'. This is how it begins:

My faither, Tam Dale, peace to his banes, was a wild, sploring lad in his young days, wi' little wisdom and less grace. He was fond of a lass and fond of a glass, and fond of a ran-dan; but I could never hear tell that he was muckle use for honest employment. Frae ae thing to anither, he listed at last for a sodger and was in the garrison of this fort... The governor brewed his ain ale; it seems it was the warst conceivable. The rock was proveesioned frae the shore with vivers, the thing was ill-guided, and there were whiles when they but to fish and shoot solans for their diet. To crown a', thir was the Days of the Persecution. The perishin' cauld chalmers were all occupeed wi' sants and martyrs, the saut of the yearth, of which it wasnae worthy. And though Tam Dale carried a firelock there, a single sodger, and liked a lass and a glass, as I was sayin', the mind of the man was mair than just set with his position. He had glints of the glory of the kirk; there were whiles when his dander rase to see the Lord's sants misguided, and shame covered him that he should be haulding a can'le (or carrying a firelock) in so black a business.

'Davie, are ye no coming?' An illustration by C E Brock from Catriona, *1928.*

RLS was very proud of his father, grandfather and uncles, and the harbours and lighthouses they built all round Scotland's coast. His father told a story of wreckers, who made a living from ships wrecked on their shores.

from A Family of Engineers

On a September night, the *Regent* lay in the Pentland Firth in a fog and a violent and windless swell. It was still dark, when they were alarmed by the sound of breakers, and an anchor was immediately let go. The peep of dawn discovered them swinging in desperate proximity to the isle of Swona and the surf bursting close under their stern. There was in this place a hamlet of the inhabitants, fisher-folk and wreckers; their huts stood close about the head of the beach. All slept; the doors were closed, and there was no smoke, and the anxious watchers on board ship seemed to contemplate a village of the dead. It was thought possible to launch a boat and tow the *Regent* from her place of danger; and with this view a signal of distress was made and a gun fired with a redhot poker from the galley. Its detonation awoke the sleepers. Door after door was opened, and in the grey light of the morning fisher after fisher was seen to come forth, yawning and stretching himself, nightcap on head. Fisher after fisher, I wrote, and my pen tripped; for it should rather stand wrecker after wrecker. There was no emotion, no animation, it scarce seemed any interest; not a hand was raised; but all callously awaited the harvest of the sea, and their children stood by their side and waited also. To the end of his life, my father remembered that amphitheatre of placid spectators on the beach, and with a special and natural animosity, the boys of his own age.

Map of Scotland. RLS visited many of the islands, lighthouses and seaside towns shown on the map.

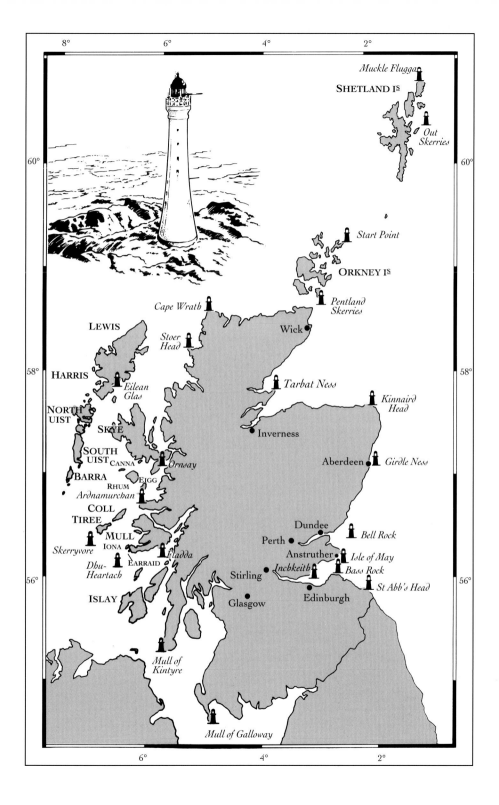

8° 6° 4° 2°

Muckle Flugga

SHETLAND Iˢ

Out Skerries

60° 60°

Start Point

ORKNEY Iˢ

Pentland Skerries

Cape Wrath Wick

Stoer Head

58° 58°

LEWIS

HARRIS

Eilean Glas

NORTH UIST

SKYE

Tarbat Ness

Kinnaird Head

Inverness

SOUTH UIST **CANNA**

Ornsay

Aberdeen *Girdle Ness*

BARRA **RHUM** EIGG

Ardnamurchan

COLL

TIREE

Skerryvore **MULL**

IONA

Dundee *Bell Rock*

Perth

ISLAY

Dhu-Heartach EARRAID *Fladda*

Anstruther *Isle of May*

Inchkeith *Bass Rock*

Stirling Edinburgh *St Abb's Head*

56° 56°

Glasgow

Mull of Kintyre

Mull of Galloway

6° 4° 2°

RLS lived for three years in Bournemouth, where he called his house 'Skerryvore', after the famous lighthouse built near Tiree in the Hebrides by his uncle Alan Stevenson.

Skerryvore

For love of lovely words, and for the sake
Of those, my kinsmen and my countrymen
Who early and late in the windy ocean toiled
To plant a star for seamen, where was then
The surfy haunt of seals and cormorants:
I, on the lintel of this cot, inscribe
The name of a strong tower.

Skerryvore: the Parallel

Here all is sunny, and when the truant gull
Skims the green level of the lawn, his wing
Dispetals roses; here the house is framed
Of kneaded brick and the plumed mountain pine,
Such clay as artists fashion and such wood
As the tree-climbing urchin breaks. But there
Eternal granite hewn from the living isle
Dowelled with brute iron, rears a tower
That from its wet foundation to its crown
Of glittering glass, stands, in the sweep of winds,
Immovable, immortal, eminent.

Originally, RLS, too, was to have been a lighthouse engineer. His father was upset when RLS told him he wanted to be a writer - 'to play at home with paper' rather than take on the demanding and sometimes dangerous work of lighthouse building.

Say not of me that weakly I declined
The labours of my sires, and fled the sea,
The towers we founded and the lamps we lit,
To play at home with paper like a child.
But rather say: *In the afternoon of time*
A strenuous family dusted from its hands
The sand of granite, and beholding far
Along the sounding coast its pyramids
And tall memorials catch the dying sun,
Smiled well content, and to this childish task
Around the fire addressed its evening hours.

Bell Rock Lighthouse, 1819, *by J M W Turner.*

As a young man RLS explored every part of Edinburgh. One of his favourite places was the top of Calton Hill. There he could look out towards Leith and across the Firth of Forth. The lighthouse ('pharos') on Inchkeith was built by his grandfather Robert Stevenson.

from Edinburgh: Picturesque Notes

Leith camps on the seaside with her forest of masts; Leith roads are full of ships at anchor; the sun picks out the white pharos upon Inchkeith Island; the Firth extends on either hand from the Ferry to the May; the towns of Fifeshire sit, each in its banks of blowing smoke, along the opposite coast; and the hills inclose the

Detail from a panoramic view of the Firth of Forth, by Lady Elton, 1823.

view, except to the farthest east, where the haze of the horizon rests upon the open sea. There lies the road to Norway: a dear road for Sir Patrick Spens and his Scots Lords; and yonder smoke on the hither side of Largo Law is Aberdour, from whence they sailed to seek a queen for Scotland...

The sight of the sea, even from a city, will bring thoughts of storm and sea disaster. The sailors' wives of Leith and the fisher-woman of Cockenzie, not sitting languorously with fans but crowding to the tail of the harbour with a shawl about their ears, may still look vainly for brave Scotsmen who will return no more, or boats that have gone on their last fishing. Since Sir Patrick sailed from Aberdour, what a multitude have gone down in the North Sea!

'A Bass Rock upon dry land'. Edinburgh from
Salisbury Crags, *by John MacWhirter.*

In the same book he describes the Castle, as if it were an island like the
Bass Rock.

In the very midst stands one of the most satisfactory crags in nature
– a Bass Rock upon dry land, rooted in a garden, shaken by passing
trains, carrying a crown of battlements and turrets, and describing
its warlike shadow over the liveliest and brightest thoroughfare of the
new town. From their smoky beehives, ten stories high, the unwashed
look down upon the open squares and gardens of the wealthy; and
gay people sunning themselves along Princes Street, with its mile of
commercial palaces all beflagged upon some great occasion, see,
across a gardened valley set with statues, where the washings of the
old town flutter in the breeze at its high windows.

RLS first saw the little island of Earraid as a teenager, when he was with his father on a lighthouse inspection trip.

from Memoirs of an Islet

The little isle of Earraid lies close in to the south-west corner of the Ross of Mull: the sound of Iona on one side, across which you may see the isle and church of Columba; the open sea to the other, where you shall be able to mark, on a clear, surfy day, the breakers running white on many sunken rocks. I first saw it, or first remembered seeing it, framed in the round bull's-eye of a cabin port, the sea lying smooth along its shores like the waters of a lake, the colourless, clear light of the early morning making plain its heathery and rocky hummocks. There stood upon it, in these days, a single rude house of uncemented stones, approached by a pier of

'The colourless, clear light of the early morning.' The island of Earraid. Elizabeth Robertson

Children on a pier.
Riddell Collection

wreckwood. It must have been very early, for it was then summer, and in summer, in that latitude, day scarcely withdraws; but even at that hour the house was making a sweet smoke of peats which came to me over the bay, and the bare-legged daughters of the cotter were wading by the pier. The same day we visited the shores of the isle in the ship's boats; rowed deep into Fiddler's Hole, sounding as we went; and having taken stock of all possible accommodation, pitched on the northern inlet as the scene of operation.

Earraid was the base for constructing the lighthouse Dhu-Heartach.

In fine weather, when by the spy-glass on the hill the sea was observed to run low upon the reef, there would be a sound of preparation in the very early morning; and before the sun had risen from behind Ben More, the tender would steam out of the bay. Over fifteen sea-miles of the great blue Atlantic rollers she ploughed her way, trailing at her tail a brace of wallowing stone-lighters. The open ocean widened upon either board, and the hills of the mainland began to go down on the horizon, before she came to her unhomely destination, and lay-to at last where the rock clapped its black head above the swell, with the tall iron barrack on its spider legs, and the truncated tower, and the cranes waving their arms, and the smoke of the engine-fire rising in the mid-sea... No other life was there but that of sea-birds, and of the sea itself, that here ran like a mill-race, and growled about the outer reef for ever, and ever and again, in the calmest weather, roared and spouted on the rock itself. Times were different upon Dhu-Heartach when it blew, and the night fell dark, and the neighbour lights of Skerryvore and Rhu-val were quenched in fog, and the men sat prisoned high up in their iron drum, that then resounded with the lashing of the sprays. Fear sat with them in their sea-beleaguered dwelling; and the colour changed in anxious faces when some greater billow struck the barrack, and its pillars quivered and sprang under the blow.

Building the
Dhu-Heartach lighthouse.

Earraid appears again in *Kidnapped*, RLS's first story about David Balfour, who is kidnapped from South Queensferry and taken on board the brig *Covenant* round the coast, only to be shipwrecked.

from Kidnapped

The reef on which we had struck was close in under the south-west end of Mull, off a little isle they call Earraid, which lay low and black upon the larboard. Sometimes the swell broke clean over us; sometimes it only ground the poor brig upon the reef, so that we could hear her beat herself to pieces; and what with the great noise of the sails, and the singing of the wind, and the flying of the spray in the moonlight, and the sense of danger, I think my head must have been partly turned, for I could scarcely understand the things I saw.

David is finally washed ashore on Earraid. He fears that his friend Alan Breck may have been drowned.

With my stepping ashore I began the most unhappy part of my adventures. It was half-past twelve in the morning, and though the wind was broken by the land, it was a cold night. I dared not sit down (for I thought I should have frozen), but took off my shoes and walked to and fro upon the sand, barefoot, and beating my breast, with infinite weariness. There was no sound of man or cattle; not a cock crew, though it was about the hour of their first waking; only the surf broke outside in the distance, which put me in mind of my perils and those of my friend. To walk by the sea at that

Rough seas off the coast of Islay. Baxter Nisbet

hour of the morning, and in a place so desert-like and lonesome, struck me with a kind of fear.

As soon as the day began I put on my shoes and climbed a hill – the ruggedest scramble I ever undertook – falling, the whole way, between big blocks of granite or leaping from one to another. When I got to the top the dawn was come. There was no sign of the brig, which must have lifted from the reef and sunk. The boat, too, was nowhere to be seen. There was never a sail upon the ocean; and in what I could see of the land, was neither house nor man.

I was afraid to think what had befallen my shipmates, and afraid to look longer at so empty a scene. What with my wet clothes and weariness, and my belly that now began to ache with hunger, I had enough to trouble me without that. So I set off eastward along the south coast, hoping to find a house where I might warm myself, and perhaps get news of those I had lost. And at the worst, I considered the sun would soon rise and dry my clothes.

Opposite: *Alan Breck is rescued by the brig* Covenant. *An illustration from* Kidnapped, *1887.*

Another shipwreck story, 'The Merry Men', is set on an island RLS calls 'Aros', but it is in fact Earraid again. The hero has come to Aros to look for the ship *Espirito Santo*, wrecked when the Spanish Armada was scattered by storms. He explores under water and finds a shoe buckle on the sea bed.

from The Merry Men

I held it in my hand, and the thought of its owner appeared before me like the presence of an actual man. His weather-beaten face, his sailor's hands, his sea-voice hoarse with singing at the capstan, the very foot that had once worn that buckle and trod so much along the swerving decks – the whole human fact of him, as a creature like myself, with hair and blood and seeing eyes, haunted me in that sunny, solitary place, not like a spectre, but like some friend whom I had basely injured. Was the great treasure-ship indeed below there, with her guns and chain and treasure, as she had sailed from Spain; her decks a garden for the sea-weed, her cabin a breeding-place for fish, soundless but for the dredging water, motionless but for the waving of the tangle upon her battlements – that old, populous sea-riding castle, now a reef in Sandag bay? Or, as I thought it likelier... was this shoe-buckle bought but the other day and worn by a man of my own period in the world's history, hearing the same news from day to day, thinking the same thoughts, praying, perhaps, in the same temple with myself? However it was, I was assailed with dreary thoughts; my uncle's words, 'the dead are down there,' echoed in my ears; and though I determined to dive once more, it was with a strong repugnance that I stepped forward to the margin of the rocks.

Shipwreck. *Detail from a painting by Ivan Constantowitsch Aivazoffski.* Christie's, London

One wet afternoon during a holiday in Braemar, RLS was playing with his stepson Lloyd. They drew a map of an island...and a story of treasure and pirates began to grow in RLS's head. It begins with Billy Bones arriving at the Admiral Benbow inn.

from Treasure Island

I remember him as if it were yesterday, as he came plodding to the inn door, his sea-chest following behind him in a hand-barrow; a tall, strong, heavy, nut-brown man; his tarry pigtail falling over the shoulders of his soiled blue coat; his hands ragged and scarred, with black, broken nails; and the sabre cut across one cheek, a dirty livid white. I remember him looking round the cove and whistling to himself as he did so, and then breaking out in that old sea-song that he sang so often afterwards:-

'Fifteen men on the dead man's chest –
Yo-ho-ho, and a bottle of rum!'

in the high, old tottering voice that seemed to have been tuned and broken at the capstan bars. Then he rapped on the door with a bit of stick like a handspike that he carried, and when my father appeared, called roughly for a glass of rum. This, when it was brought to him, he drank slowly, like a connoisseur, lingering on the taste, and still looking about him at the cliffs and up at our signboard.

'This is a handy cove,' says he, at length; 'and a pleasant sittyated grog-shop. Much company, mate?'

Long John Silver examines the map of Treasure Island. An illustration by Wal Paget from Treasure Island, *1899.*

Jim Hawkins, the story's hero, finds a map in Billy Bones's sea chest, which leads him and his friends to set sail on the *Hispaniola*. At last, they reach the island marked on the map.

The appearance of the island when I came on deck next morning was altogether changed. Although the breeze had now utterly ceased, we had made a great deal of way during the night, and were now lying becalmed about half a mile to the south-east of the low eastern coast. Grey-coloured woods covered a large part of the surface. This even tint was indeed broken up by streaks of yellow sandbreak in the lower lands, and by many tall trees of the pine family, out-topping the others – some singly, some in clumps; but the general colouring was uniform and sad. The hills ran up clear above the vegetation in spires of naked rock. All were strangely shaped, and the Spy-glass, which was by three or four hundred feet the tallest on the island, was likewise the strangest in configuration, running up sheer from almost every side, then suddenly cut off at the top like a pedestal to put a statue on.

The *Hispaniola* was rolling scuppers under in the ocean swell. The booms were tearing at the blocks, the rudder was banging to and fro, and the whole ship creaking, groaning, and jumping like a manufactory. I had to cling tight to the backstay, and the world turned giddily before my eyes; for though I was a good enough sailor when there was way on, this standing still and being rolled about like a bottle was a thing I never learned to stand without a qualm or so, above all in the morning, on an empty stomach.

Perhaps it was this – perhaps it was the look of the island, with its grey, melancholy woods, and wild stone spires, and the surf that

Map of Treasure Island, drawn by RLS for the first edition of his book - but not the original map he drew for Lloyd in Braemar.

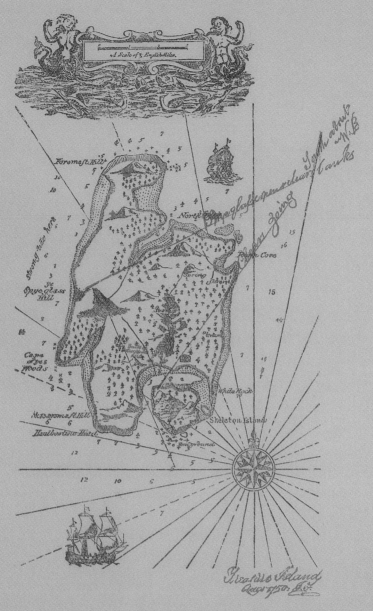

A Scale of 3 English Miles.

Foremast Hill

North Inlet

The Spye glass quarduin Hawks being N.E.

Spy glass

John being

Strong tide here

The Spye glass Hill

Cape of the Woods

Maingrove of the Hill

Haulbowline Head

Spring

Wood

Mizzen

Hinders

Cove

White Rock

Skeleton Island

Bury ground

Treasure Island
Augst 1750. W.B.

Given above S.F. & M. W. Bones Maate of ye Walrus
Savannah this twenty Saly 1754 W. B.

Facsimile of Chart, latitude and
Longitude struck out by J. Hawkins

we could both see and hear foaming and thundering on the steep beach – at least, although the sun shone bright and hot, and the shore birds were fishing and crying all around us, and you would have thought anyone would have been glad to get to land after being so long at sea, my heart sank, as the saying is, into my boots; and from that first look onward, I hated the very thought of Treasure Island.

After many adventures, battles and dangers, Jim and his companions find the treasure, and prepare to leave the island.

That was about our last doing on the island. Before that, we had got the treasure stowed, and had shipped enough water and the remainder of the goat meat, in case of any distress; and at last, one fine morning, we weighed anchor, which was about all that we could manage, and stood out of North Inlet, the same colours flying that the captain had flown and fought under at the palisade.

The three fellows must have been watching us closer than we thought for, as we soon had proved. For, coming through the narrows, we had to lie very near the southern point, and there we saw all three of them kneeling together on a spit of sand, with their arms raised in supplication. It went to all our hearts, I think, to leave them in that wretched state; but we could not risk another mutiny; and to take them home for the gibbet would have been a cruel sort of kindness. The doctor hailed them and told them of the stores we had left, and where they were to find them. But they continued to call us by name, and appeal to us, for God's sake, to be merciful, and not leave them to die in such a place.

The Hispaniola *leaves the island and three buccaneers 'their arms raised in supplication'. An illustration by Wal Paget from* Treasure Island, *1899.*

At last, seeing the ship still bore on her course, and was now swiftly drawing out of earshot, one of them – I know not which it was – leapt to his feet with a hoarse cry, whipped his musket to his shoulder, and sent a shot whistling over Silver's head and through the mainsail.

After that, we kept under cover of the bulwarks, and when next I looked out they had disappeared from the spit, and the spit itself had almost melted out of sight in the growing distance. That was, at least, the end of that; and before noon, to my inexpressible joy, the highest rock of Treasure Island had sunk into the blue round of sea.

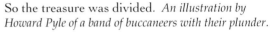

So the treasure was divided. *An illustration by Howard Pyle of a band of buccaneers with their plunder.*

RLS and his family set sail from San Francisco in 1888. He loved visiting the Pacific islands and meeting their people but he often thought of Scotland. In a letter to his friend Charles Baxter he told how he had thought of Edinburgh while sailing through a very dangerous part of the Pacific. 'Rutherfords' is the pub he frequented as a student.

from *Letter to Charles Baxter*

Last night as I lay under my blanket in the cockpit, courting sleep, I had a comic seizure. There was nothing visible but the southern stars, and the steersman there out by the binnacle lamp; we were all looking forward to a most deplorable landfall on the morrow, praying God we should fetch a tuft of palms which are to indicate the Dangerous Archipelago; the night was as warm as milk, and all of a sudden I had a vision of – Drummond Street. It came on me like a flash of lightning: I simply returned thither, and into the past. And when I remember all I hoped and feared as I pickled about Rutherford's in the rain and the east wind; how I feared I should make a mere shipwreck, and yet timidly hoped not; how I feared I should never have a friend, far less a wife, and yet passionately hoped I might; how I hoped (if I did not take to drink) I should possibly write one little book, etc. etc. And then now – what a change! I feel somehow as if I should like the incident set upon a brass plate at the corner of that dreary thoroughfare for all students to read, poor devils, when their hearts are down.

Map of the South Pacific showing RLS's voyages on the Casco, *the* Equator *and the* Janet Nichol.

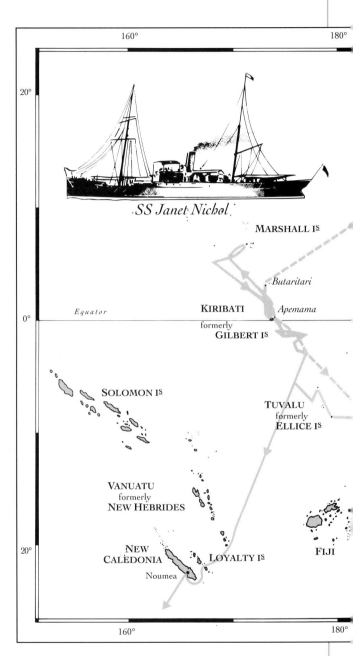

SS Janet Nichol

MARSHALL IS

Butaritari

Equator

KIRIBATI   Apemama

formerly
GILBERT IS

SOLOMON IS

TUVALU
formerly
ELLICE IS

VANUATU
formerly
NEW HEBRIDES

NEW
CALEDONIA   LOYALTY IS

FIJI

Noumea

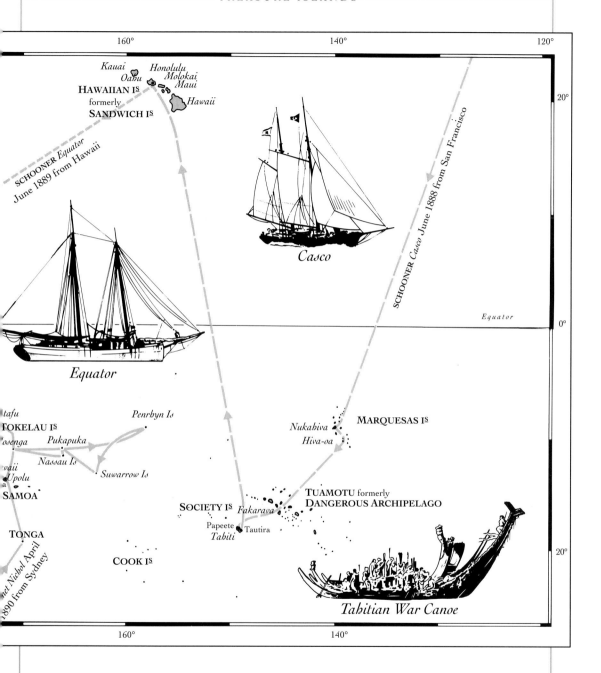

160°   140°   120°

Kauai
Oahu   *Honolulu*
HAWAIIAN Iˢ   *Molokai*
formerly   *Maui*
SANDWICH Iˢ   *Hawaii*

20°

SCHOONER *Equator*
June 1889 from Hawaii

SCHOONER *Casco* June 1888 from San Francisco

Casco

Equator   0°

Equator

tafu
TOKELAU Iˢ   *Penrhyn Is*
osenga   *Pukapuka*
Nukahiva   **MARQUESAS I**ˢ
Nassau Is
vaii   *Hiva-oa*
Upolu
a   *Suwarrow Is*
SAMOA

TUAMOTU formerly
SOCIETY Iˢ   *Fakarava*   **DANGEROUS ARCHIPELAGO**

TONGA
Papeete   Tautira
Tahiti

t Nichol April
890 from Sydney
COOK Iˢ

20°

Tahitian War Canoe

160°   140°

The first islands the Stevenson family came to were the Marquesas. There was great excitement when the voyagers saw land appear.

from In the South Seas

I turned shoreward, and high squalls were overhead; the mountains loomed up black; and I could have fancied I had slipped ten thousand miles away and was anchored in a Highland loch; that when the day came, it would show pine, and heather, and green fern, and roofs of turf sending up the smoke of peats; and the alien speech that should next greet my ears must be Gaelic, not Kanaka.

... I have watched the morning break in many quarters of the world; it has been certainly one of the chief joys of my existence, and the dawn that I saw with most emotion shone upon the bay of Anaho. The mountains abruptly overhang the port with every variety of surface and of inclination, lawn, and cliff, and forest. Not one of these but wore its proper tint of saffron, of sulphur, of the clove, and of the rose. The lustre was like that of satin; on the lighter hues there seemed to float an efflorescence; a solemn bloom appeared on the more dark. The light itself was the ordinary light of morning, colourless and clean; and on this ground of jewels, pencilled out the least detail of drawing. Meanwhile, around the hamlet, under the palms, where the blue shadow lingered, the red coals of cocoa husk and the light trails of smoke betrayed the awakening business of the day; along the beach men and women, lads and lasses, were returning from the bath in bright raiment, red and blue and green, such as we delighted to see in the coloured pictures of our childhood;and presently the sun had cleared the eastern hill, and the glow of the day was over all.

A Tahitian family on the seashore, by Paul Gauguin, 1902.

RLS spent hours exploring the shore, much as he had done as a child in Scotland.

My favourite haunt was opposite the hamlet, where there was a landing in a cove under a lianaed cliff. The beach was lined with palms and a tree called the purao, something between a fig and a mulberry in growth, and bearing a flower like a great yellow poppy with a maroon heart. In places rocks encroached upon the sand; the beach would be all submerged; and the surf would bubble warmly as high as to my knees, and play with cocoa-nut husks as our more homely ocean plays with wreck and wrack and bottles. As the reflux drew down, marvels of colour and design streamed between my feet; which I would grasp at, miss, or seize: now to find them what they promised, shells to grace a cabinet or be set in gold upon a lady's finger; now to catch only *maya* of coloured sand, pounded fragments and pebbles, that, as soon as they were dry, became as dull and homely as flints upon a garden path. I have toiled at this childish pleasure for hours in the strong sun, conscious of my incurable ignorance; but too keenly pleased to be ashamed. Meanwhile, the blackbird (or his tropical understudy) would be fluting in the thickness overhead.

A little further, in the turn of the bay, a streamlet trickled in the bottom of a den, thence spilling down a stair of rock into the sea. The draught of air drew down under the foliage in the very bottom of the den, which was a perfect arbour for coolness. In front it stood open on the blue bay and the *Casco* lying there under her awning and her cheerful colours. Overhead was a thatch of puraos, and over these again palms brandished their bright fans, as I have seen a conjurer make himself a halo out of naked swords. For in this spot, over a neck of low land at the foot of the mountains, the trade-wind streams into Anaho Bay in a flood of almost constant volume and velocity, and of heavenly coolness.

'She is a lovely creature' - the schooner yacht Casco.

RLS felt that he had a lot in common with the Pacific islanders. They were not so different from the islanders he knew at home.

These points of similarity between a South Sea people and some of my own folk at home ran much in my head in the islands; and not only inclined me to view my fresh acquaintances with favour, but continually modified my judgment. A polite Englishman comes today to the Marquesans and is amazed to find the men tattooed; polite Italians came not long ago to England and found our fathers stained with woad; and when I paid the return visit as a little boy, I was highly diverted with the backwardness of Italy: so insecure, so much a matter of the day and hour, is the pre-eminence of race.

A Marquesan family. The man is tattooed all over his body.
From A Trip Round the World, *by A St Aulaire.*

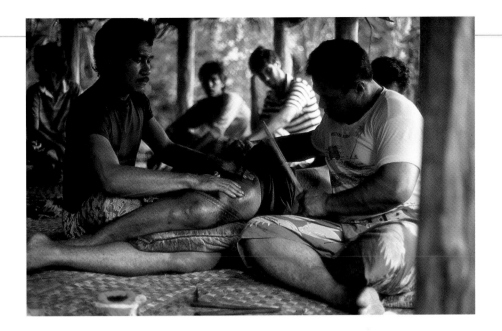

Modern tattooing, in Western Samoa. Merrily Stover

War clubs from the Marquesas and New Caledonia. The Marquesan clubs are like those in the picture on the previous page.

Stevenson wrote *The Wrecker* with his step-son Lloyd Osbourne. The story begins in the Marquesas Islands. A strange ship approaches the port of Tai-o-hae, and the first to spot it is a tattooed European who has spent many years in the Pacific. (The word 'Kanaka' sometimes used to mean Polynesian is Hawaiian for 'man'.)

from The Wrecker

His eyes were open, staring down the bay. He saw the mountains droop, as they approached the entrance, and break down in cliffs: the surf boil white round the two sentinel islets; and between, on the narrow bight of blue horizon, Ua-pu upraised the ghost of her pinnacled mountain-tops. But his mind would take no account of these familiar features; as he dodged in and out along the frontier line of sleep and waking, memory would serve him with broken fragments of the past: brown faces and white, of skipper and ship-mate, king and chief, would arise before his mind and vanish; he would recall old voyages, old landfalls in the hour of dawn; he would hear again the drums beat for a man-eating festival; perhaps he would summon up the form of that island princess for the love of whom he had submitted his body to the cruel hands of the tattooer, and now sat on the lumber, at the pier-end of Tai-o-hae, so strange a figure of a European. Or perhaps, from yet further back, sounds and scents of England and his childhood might assail him: the merry clamour of cathedral bells, the broom upon the foreland, the song of the river on the weir.

It is bold water at the mouth of the bay; you can steer a ship about either sentinel, close enough to toss a biscuit on the rocks. Thus it chanced that, as the tattooed man sat dozing and dreaming, he was startled into wakefulness and animation by the appearance of a flying jib beyond the western islet. Two more headsails followed; and before the tattooed man had scrambled to his feet, a topsail schooner, of some hundred tons, had luffed about the sentinel, and was standing up the bay, close-hauled.

The sleeping city awakened by enchantment. Natives appeared upon all sides, hailing each other with the magic cry 'Ehippy' - ship; the Queen stepped forth on her verandah, shading her eyes under a hand that was a miracle of the fine art of tattooing; the commandant broke from his domestic convicts and ran into the residency for his glass; the harbour master, who was also the gaoler, came speeding down the Prison Hill; the seventeen brown Kanakas and the French boatswain's mate, that made up the complement of the war-schooner, crowded on the forward deck; and the various English, Americans, Germans, Poles, Corsicans, and Scots - the merchants and the clerks of Tai-o-hae – deserted their places of business, and gathered, according to invariable custom, on the road before the club.

So quickly did these dozen whites collect, so short are the distances in Tai-o-hae, that they were already exchanging guesses as to the nationality and business of the strange vessel, before she had gone about upon her second board towards the anchorage. A moment after, English colours were broken out at the main truck.

Children today in Kiribati (the Gilbert Islands).
James Siers

Children in Western Samoa, on the way to church. Merrily Stover

From the Marquesas the schooner *Casco* with the Stevensons on board sailed on through the Dangerous Archipelago (now called Tuamotu) on to Tahiti. From there RLS wrote to Tom Archer, young son of his friend William Archer.

from Letter to Tom Archer

This is a much better place for children than any I have hitherto seen in these seas. The girls (and sometimes the boys) play a very elaborate kind of hopscotch. The boys play horses exactly as we do in Europe; and have very good fun on stilts, trying to knock each other down, in which they do not often succeed. The children of all ages go to church and are allowed to do what they please, running about the aisles, rolling balls, stealing mama's bonnet and publicly sitting on it, and at last going to sleep in the middle of the floor. I forgot to say that the whips to play horses, and the balls to roll about the church – at least I never saw them used elsewhere – grow ready made on trees; which is rough on toy-shops. The whips are so good that I wanted to play horses myself; but no such luck! my hair is grey, and I am a great big, ugly man.

In Tahiti RLS became great friends with Prince Ori à Ori and his sister Princess Moë.

from To an Island Princess

Since long ago, a child at home,
I read and longed to rise and roam,
Where'er I went, whate'er I willed,
One promised land my fancy filled.
Hence the long roads my home I made;
Tossed much in ships; have often laid
Below the uncurtained sky my head,
Rain-deluged and wind-buffeted:
And many a thousand hills I crossed
And corners turned – Love's labour lost,
Till, Lady, to your isle of sun
I came, not hoping; and, like one
Snatched out of blindness, rubbed my eyes,
And hailed my promised land with cries.

RLS sailed away from Tahiti on Christmas Day 1888, and arrived in the Sandwich Islands – Hawaii – in the New Year. He wrote a letter about the voyage to his cousin Bob.

from Letter to Bob Stevenson

RLS and crew members on the bowsprit of the Equator.

One stirring day was that in which we sighted Hawaii. It blew fair, but very strong; we carried jib, foresail, and mainsail, all single-reefed, and she carried her lee rail under water and flew. The swell, the heaviest I have ever been out in – I tried in vain to estimate the height, at least fifteen feet – came tearing after us about a point and a half off the wind. We had the best hand – old Louis – at the wheel; and really, he did nobly, and had noble luck, for it never caught us once. At times it seemed we must have it; Louis would look over his shoulder with the queerest look and dive down his neck into his shoulders; and then it missed us somehow, and only sprays came over our quarter, turning the little outside lane of deck into a mill race as deep as to the cockpit coamings. I never remember anything more delightful and exciting. Pretty soon after we were lying absolutely becalmed under the lee of Hawaii, of which we had been warned; and the captain never confessed he had done it on purpose, but when accused, he smiled.

*During his Pacific travels RLS collected objects like
these. The tapa cloth (made from bark) in the back-
ground came from his Samoan home, Vailima.*

Later on, he wrote to Adelaide Boodle, who had been a neighbour in Bournemouth. By that time he had decided that Honolulu was too up-to-date for him, with its electric lights and telephones. He was longing to get back to sea.

from Letter to Adelaide Boodle

The Sandwich Islands do not interest us very much; we live here, oppressed with civilisation, and look for good things in the future. But it would surprise you if you came out tonight from Honolulu (all shining with electric lights, and all in a bustle from the arrival of the mail, which is to carry you these lines) and crossed the long wooden causeway along the beach, and came out on the road through Kapiolani park, and seeing a gate in the palings, with a tub of gold-fish by the wayside, entered casually in. The buildings stand in three groups by the edge of the beach, where an angry little spitfire sea continually spirts and thrashes with impotent irascibility, the big seas breaking further out upon the reef. The first is a small house, with a very large summer parlour, or lanai, as they call it here, roofed, but practically open. There you will find the lamps burning and the family sitting about the table, dinner just done: my mother, my wife, Lloyd, Belle, my wife's daughter, Austin her child, and tonight (by way of rarity) a guest. All about the walls our South Sea curiosities, war clubs, idols, pearl shells, stone axes, etc.; and the walls are only a small part of a lanai, the rest being glazed or latticed windows, or mere open space.

During his stay in Hawaii, RLS met Princess Kaiulani, the king's niece, and half Polynesian, half Scottish. He wrote this poem to her, when she was about to depart for Scotland to complete her education.

To *Princess Kaiulani*

Princess Kaiulani.

Forth from her land to mine she goes
The island maid, the island rose,
Light of heart and bright of face:
The daughter of a double race.

Her islands here, in Southern sun,
Shall mourn their Kaiulani gone,
And I, in her dear banyan shade,
Look vainly for my little maid.

But our Scots islands far away
Shall glitter with unwonted day,
And cast for once their tempests by
To smile in Kaiulani's eye.

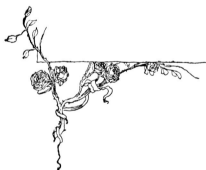

He also wrote a poem for the king, Kalakaua, who became a great friend. The *Casco* was called 'the silver ship' by the Polynesians.

To Kalakaua

The Silver Ship, my King – that was her name
In the bright islands whence your fathers came –
The Silver Ship, at rest from winds and tides,
Below your palace in the harbour rides:
And the seafarers, sitting safe on shore,
Like eager merchants count their treasures o'er.
One gift they find, one strange and lovely thing,
Now doubly precious since it pleased a king.

The right, my liege, is ancient as the lyre
For bards to give to kings what kings admire.
'Tis mine to offer for Apollo's sake;
And since the gift is fitting, yours to take.
To golden hands the golden pearl I bring:
The ocean jewel to the island king.

RLS, Lloyd and King Kalakaua of Hawaii.

The voyages RLS made were often dangerous, and there were several narrow escapes. There were squalls and storms, and reefs not always marked on charts.

from Letter to Charles Baxter

This new cruise of ours is somewhat venturesome; and I think it needful to warn you not to be in a hurry to suppose us dead. In these ill-charted seas, it is quite on the cards we might be cast on some unvisited, or very rarely visited, island; that there we might lie for a long time, even years, unheard of; and yet turn up smiling at the hinder end. So do not let me be 'rowpit' till you get some certainty we have gone to Davie Jones in a squall, or graced the feast of some barbarian in the character of Long Pig.

from Letter to Sidney Colvin

Rain, calms, squalls, bang – there's the foreto mast gone; rain, calm, squalls, away with the stay-sail; more rain, more calm, more squalls; a prodigious heavy sea all the time, and the *Equator* staggering and hovering like a swallow in a storm; and the cabin a great square, crowded with wet human beings, and the rain avalanching the deck, and the leaks dripping everywhere: Fanny in the midst of fifteen males, bearing up wonderfully.

Detail from A schooner driven towards rocks, *by* James Wilson Carmichael. Bonham's, London

RLS wrote several stories and poems based on Polynesian tales. In his story, 'The Isle of Voices', Keola is trying to escape from Kalamake, a magician. Though picked up from the sea by a ship, Keola is tormented by the mate, and decides to run away.

from The Isle of Voices

It was a fine starry night, the sea was smooth as well as the sky fair; it blew a steady trade; and there was the island on their weather bow, a ribbon of palm-trees lying flat along the sea. The captain and the mate looked at it with the night-glass, and named the name of it, and talked of it, beside the wheel where Keola was steering. It seemed it was an isle where no traders came. By the captain's way, it was an isle besides where no man dwelt; but the mate thought otherwise.

'I don't give a cent for the directory,' he said. 'I've been past here one night in the schooner *Eugenie*; it was just such a night as this; they were fishing with torches, and the beach was thick with lights like a town.'

'Well, well,' says the captain, 'it's steep-to, that's the great point; and there ain't any outlying dangers by the chart, so we'll just hug the side of it. – Keep her romping full, don't I tell you!' he cried to Keola, who was listening so hard that he forgot to steer.

And the mate cursed him, and swore that Kanaka was for no use in the world, and if he got started after him with a belaying pin, it would be a cold day for Keola.

And so the captain and the mate lay down on the house together, and Keola was left to himself.

'This island will do very well for me,' he thought; 'if no traders deal there, the mate will never come. And as for Kalamake, it is not possible he can ever get as far as this.'

With that he kept edging the schooner nearer in. He had to do this quietly, for it was the trouble with these white men, and above all with the mate, that you could never be sure of them; they would all be sleeping sound, or else pretending, and if a sail shook they would jump to their feet and fall on you with a rope's end. So Keola edged her up little by little, and kept all drawing. And presently the land was close on board, and the sound of the sea on the sides of it grew loud.

With that the mate sat up suddenly upon the house.

'What are you doing? he roars. 'You'll have the ship ashore!'

And he made one bound for Keola, and Keola made another clean over the rail and plump into the starry sea. When he came up again, the schooner had paid off on her true course, and the mate stood by the wheel himself, and Keola heard him cursing. The sea was smooth under the lea of the island; it was warm besides, and Keola had his sailor's knife, so he had no fear of sharks. A little way before him the trees stopped; there was a break in the line of the land like the mouth of a harbour; and the tide, which was then flowing, took him up and carried him through. One minute he was without, and the next within; had floated there in a wide shallow water, bright with ten thousand stars, and all about him was the ring of the land, with its string of palm-trees. And he was amazed, because this was a kind of island he had never heard of.

RLS's long poem 'Rahéro' is an English version of a Tahitian story of vengeance. In this passage the people of one Tahitian village set out to visit another, not knowing that a trap has been laid.

from Rahéro

Soon as the morning broke,
Canoes were thrust in the sea, and the houses emptied
of folk.
Strong blew the wind of the south, the wind that gathers
the clan;
Along all the line of the reef the clamorous surges ran;
And the clouds were piled on the top of the island
mountain-high,
A mountain throned on a mountain. The fleet of canoes
swept by
In the midst, on the green lagoon, with a crew released
from care,
Sailing an even water, breathing a summer air,
Cheered by a cloudless sun; and never to left and right,
Bursting surge on the reef, drenching storms on the
height.
So the folk of Vaiau sailed and were glad all day,
Coasting the palm-tree cape and crossing the populous bay
By all the towns of the Tevas; and still as they bowled
along,
Boat would answer boat with jest and laughter and song,

And the people of all the towns trooped to the sides of
the sea
And gazed from under the hand or sprang aloft on the
tree,
Hailing and cheering. Time failed them for more to do;
The holiday village careened to the wind, and was gone
from view
Swift as a passing bird; and ever as onward it bore,
Like the cry of the passing bird, bequeathed its song to
the shore -
Desirable laughter of maids and the cry of delight of the
child.
And the gazer, left behind, stared at the wake and smiled.
By all the towns of the Tevas they went, and Pápara last,

Tahitian war canoes at Pari, by William Hodges.

The home of the chief, the place of muster of war; and
passed
The march of the lands of the clan, to the lands of an
alien folk.
And there, from the dusk of the shoreside palms, a column
of smoke
Mounted and wavered and died in the gold of the setting
sun,
'Paea!' they cried. 'It is Paea.' And so was the voyage
done.

In the early fall of the night, Hiopa came to the shore,
And beheld and counted the comers, and lo, they were
forty score:
The pelting feet of the babes than ran already and
played,
The clean-lipped smile of the boy, the slender breasts of
the maid,
And mighty limbs of women, stalwart mothers of men.
...
But these were foes of his clan,
And he trod upon pity, and came, and civilly greeted
the king,
And gravely entreated Rahéro; and for all that could
fight or sing,
And claimed a name in the land, had fitting phrases of
praise;
And for all who were well-descended he spoke of the
ancient of days.

RLS left Hawaii in June 1889, hoping to reach the Gilbert Islands (now called Kiribati), coral atolls low on the ocean. They were sailing in a more remote part of the Pacific than before.

from *In the South Seas*

The whole extent of the South Seas is a desert of ships; more especially that part where we were now to sail. No post runs in these islands; communication is by accident; where you may have designed to go is one thing, where you shall be able to arrive, another. It was my hope, for instance, to have reached the Carolines, and returned to the light of day by way of Manila and the China ports; and it was in Samoa that we were destined to reappear and be once more refreshed with the sight of mountains. Since the sunset faded from the peaks of Oahu six months had intervened, and we had seen no spot of earth so high as an ordinary cottage. Our path had been still on the flat sea, our dwellings upon unerected coral, our diet from the pickle-tub or out of tins; I had learned to welcome shark's flesh for a variety; and a mountain, an onion, an Irish potato or a beefsteak, had been long lost to sense and dear to aspiration.

A coral atoll on a flat sea.

RLS realized that his health would suffer badly if he left the tropics. He decided to buy land and build a house in Samoa. In August 1890 he was in Sydney, Australia, and wrote a letter to his old friend, the writer Henry James.

from Letter to Henry James

I do not think I shall come to England more than once, and then it'll be to die. Health I enjoy in the tropics; even here, which they call sub- or semi-tropical, I come only to catch cold. I have not been out since my arrival; live here in a nice bedroom by the fireside, and read books and letters from Henry James... But I can't go out! The thermometer was down to nearly 50° the other day - no temperature for me, Mr. James: how should I do in England! I fear not at all. Am I very sorry? I am sorry about seven or eight people in England, and one or two in the States. And outside of that, I simply prefer Samoa. These are the words of honesty and soberness. (I am fasting from all but sin, coughing, *The Bondman*, a couple of eggs and a cup of tea.) I was never fond of towns, houses, society, or (it seems) civilisation. Nor yet it seems was I ever fond of (what is technically called) God's green earth. The sea, the islands, the islanders, the island life and climate, make and keep me truly happier. These last two years I have been much at sea, and I have *never wearied*; sometimes I have indeed grown impatient for some destination; more often I was sorry that the voyage drew so early to an end; and never once did I lose my fidelity to blue water and a ship. It is plain, then, that for me my exile to the place of schooners and islands can be in no sense regarded as a calamity.

Detail from Te Raau Rahi
(The big tree), by Paul Gauguin.

RLS was ill in Sydney and left on the steamer *Janet Nichol*. It proved an adventurous voyage.

from Letter to Sidney Colvin

We left Sydney, had a cruel rough passage to Auckland, for the *Janet* is the worst roller I was ever aboard of. I was confined to my cabin, ports closed, self shied out of the berth, stomach (pampered till the day I left on a diet of perpetual egg-nogg) revolted at the ship's food and ship eating, in a frowsy bunk, clinging with one hand to the plate, with the other to the glass, and using the knife and fork (except at intervals) with the eyelid. No matter: I picked up hand over hand. After a day in Auckland, we set sail again; were blown up in the main cabin with calcium fires, as we left the bay. Let no man say I am unscientific: when I ran, on the alert, out of my stateroom, and found the main cabin incarnadined with the glow of the last scene of a pantomime, I stopped dead: 'What is this?' said I. 'This ship is on fire, I see that; but why a pantomime?' And I stood and reasoned the point, until my head was so muddled with the fumes that I could not find the companion. A few seconds later the captain had to enter crawling on his belly, and took days to recover (if he has recovered) from the fumes. By singular good fortune, we got the hose down in time and saved the ship, but Lloyd lost most of his clothes and a great part of our photographs was destroyed. Fanny saw the native sailors tossing overboard a blazing trunk; she stopped them in time, and behold, it contained my manuscripts.

RLS settled in Samoa in 1890. Looking after his house and land was hard work, and he was also busily writing - stories, articles, poems and many letters. He wrote this letter to entertain children in the care of Miss B...

from Letter to Miss B...

This man lives on an island which is not very long and is extremely narrow. The sea beats round it very hard, so that it is difficult to get to shore. There is only one harbour where ships come, and even that is very wild and dangerous; four ships of war were broken there a little while ago, and one of them is still lying on its side on a rock clean above water, where the sea threw it as you might throw your fiddle-bow upon the table. All round the harbour the town is strung out: it is nothing but wooden houses, only there are some churches built of stone. They are not very large, but the people have

A shipwreck, *by Alexander Nasmyth.*

Island houses in Western Samoa. Merrily Stover

never seen such fine buildings. Almost all the houses are of one stor[e]y. Away at one end of the village lives the king of the whole country. His palace has a thatched roof which rests upon posts; there are no walls, but when it blows and rains, they have Venetian blinds which they let down between the posts, making all very snug. There is no furniture, and the king and the queen and the courtiers sit and eat on the floor, which is of gravel: the lamp stands there too, and every now and then it is upset...

The road to this lean man's house is uphill all the way, and through forests; the trees are not so much unlike those at home, only here and there some very queer ones are mixed with them - cocoanut palms, and great trees that are covered with bloom like red hawthorn but not near so bright; and from them all thick creepers hang down like ropes, and ugly-looking weeds that they call orchids grow in the forks of the branches; and on the ground many prickly things are dotted, which they call pine-apples. I suppose every one has eaten pine-apple drops.

On the way up to the lean man's house you pass a little village, all of houses like the king's house, so that as you ride by you can see

The gate at the entrance to Vailima. James Siers

everybody sitting at dinner, or, if it is night, lying in their beds by lamplight; because all the people are terribly afraid of ghosts, and would not lie in the dark for anything. After the village, there is only one more house, and that is the lean man's. For the people are not very many, and live all by the sea, and the whole inside of the island is desert woods and mountains. When the lean man goes into the forest, he is very much ashamed to own it, but he is always in a terrible fright. The wood is great, and empty, and hot, and it is always filled with curious noises: birds cry like children, and bark like dogs...

In a letter to his cousin Bob, RLS
described what life was like at Vailima.
He still had to be careful of his health.
Sometimes when he was tired or not
well he dictated his work to his step-
daughter Belle.

from Letter to Bob Stevenson

I have a room now, a part of the twelve-foot verandah sparred in,
at the most inaccessible end of the house. Daily I see the sunrise
out of my bed, which I still value as a tonic, a perpetual tuning fork,
a look of God's face once in the day. At six my breakfast comes up
to me here, and I work until eleven. If I am quite well, I sometimes
go out and bathe in the river before lunch, twelve. In the afternoon
I generally work again, now with Belle dictating. Dinner is at six,
and I am often in bed by eight. This supposing me to stay at home.
But I must often be away, sometimes all day long, sometimes till
twelve, one, or two at night, when you might see me coming home to
the sleeping house, sometimes in a trackless darkness, sometimes with
a glorious tropic moon, everything drenched with dew - unsaddling
and creeping to bed; and you would no longer be surprised that I live
out in this country, and not in Bournemouth – in bed.

The Ebb-Tide was the last complete novel RLS wrote. It is about two Englishmen and an American on the run, illegally sailing a schooner and looking for somewhere to hide. They approach an island which they hope is uninhabited.

from The Ebb-Tide

Meanwhile the captain was in the four cross-trees, glass in hand, his eyes in every quarter, spying for an entrance, spying for signs of tenancy. But the isle continued to unfold itself in joints, and to run out in indeterminate capes, and still there was neither house nor man, nor smoke of fire. Here a multitude of sea-birds soared and twinkled, and fished in the blue waters; and there, and for miles together, the fringe of cocoapalm and pandanus extended desolate, and made desirable green bowers for nobody to visit, and the silence of death was only broken by the throbbing of the sea.

The airs were very light, their speed was small; the heat intense. The decks were scorching underfoot, the sun flamed overhead, brazen, out of a brazen sky; the pitch bubbled in the seams, and the brains in the brainpan. And all the while the excitement of the three adventurers glowed about their bones like a fever. They whispered, and nodded, and pointed, and put mouth to ear, with a singular instinct of secrecy, approaching that island under-hand like eavesdroppers and thieves; and even Davis from the cross-trees gave his order mostly by gestures. The hands shared in this mute strain, like dogs, without comprehending it; and through the roar of so many miles of breakers, it was a silent ship that approached an empty island.

'the fringe of cocoapalm...' Jean-Paul Nacivet

They land on the island, and find it inhabited after all. Herrick, the book's hero, begins to explore.

When he now went forward it was cool with the shadow of many well-grown palms; draughts of the dying breeze swung them together overhead; and on all sides, with a swiftness beyond dragon-flies or swallows, the spots of sunshine flitted, and hovered, and returned. Underfoot, the sand was fairly solid and quite level, and Herrick's steps fell there noiseless as in new-fallen snow. It bore the marks of having been once weeded like a garden alley at home; but the pestilence had done its work, and the weeds were returning. The buildings of the settlement showed here and there through the stems of the colonnade, fresh painted, trim and dandy, and all silent scurry and some crowing of poultry; and from behind the house with the verandahs he saw smoke arise and heard the crackling of a fire.

The stone houses were nearest him upon his right. The first was locked; in the second he could dimly perceive, through a window, a certain accumulation of pearl-shell piled in the far end; the third, which stood gaping open on the afternoon, seized on the mind of Herrick with its multiplicity and disorder of romantic things. Therein were cables, windlasses and blocks of every size and capacity; cabin windows and ladders; rusty tanks, a companion hatch; a binnacle with its brass mountings and its compass idly pointing, in the confusion and dusk of that shed, to a forgotten pole; ropes, anchors, harpoons, a blubber-dipper of copper, green with years, a steering-wheel, a tool-chest with the vessel's name upon the top, the *Asia*: a whole curiosity-shop of sea-curios, gross and solid, heavy to lift, ill to break, bound with brass and shod with iron.

Two wrecks at the least must have contributed to this random heap of lumber; and as Herrick looked upon it, it seemed to him as if the two ships' companies were there on guard, and he heard the tread of feet and whisperings, and saw with the tail of his eye the commonplace ghosts of sailor men.

This was not merely the work of an aroused imagination, but had something sensible to go upon; sounds of a stealthy approach were no doubt audible; and while he still stood staring at the lumber, the voice of his host sounded suddenly, and with even more than the customary softness of enunciation, from behind.

'Junk,' it said, 'only old junk!'

A 'whole curiosity shop of sea curios'. From the collections of the National Museums of Scotland.

The *'hills of home'*. Summer Moorfoot, *by William Gillies.*

RLS died suddenly, one evening at Vailima when he and Fanny were preparing supper. Up to the day he died he was still writing about his first island country, Scotland. This poem about Scotland was written in 1893, the year before he died.

To S R Crockett

Blows the wind today, and the sun and the wind
                                        are flying,
Blows the wind on the moors today and now,
Where about the graves of the martyrs the whaups
                                        are crying,
My heart remembers how.

Grey recumbent tombs of the dead in desert places,
Standing-stones on the vacant wine-red moor,
Hills of sheep, and the howes of the silent vanished
                                        races,
And winds, austere and pure;

Be it granted to me to behold you again in
                                        dying,
Hills of home! and to hear again the call;
Hear about the graves of the martyrs the peewees
                                        crying,
And hear no more at all.

The passages in this anthology are taken from the following books by Robert Louis Stevenson: